The AI-Enabled Leader

A Human First Playbook for Building Trust, Capability & Results with AI

By Beth Barratt-Browne

2025

www.BetterHRx.com.au

First edition 2025

ISBN: 978-1-7642856-3-6 *paperback*
ISBN: 978-1-7642856-2-9 *ebook*

Cover and interior by 4HG Design | www.4HG.com.au

Abstract

This book provides a comprehensive guide to integrating artificial intelligence into your organisation while maintaining the critical human element that drives workplace success. As AI transforms business operations across industries, leaders face the challenge of balancing technological advancement with human-centered workplace cultures. The Human Element offers a strategic roadmap for this journey, combining practical frameworks, ethical considerations, and leadership insights to help you build effective AI-human partnerships. Rather than focusing on technology alone, we explore the organisational, cultural, and human dimensions essential for successful digital transformation. Through research-backed strategies and real-world case studies, you'll discover how to assess your organisation's AI readiness, develop new hybrid team structures, and implement governance models that foster innovation while preserving your company's values and workforce strengths. Whether you're a CEO contemplating your first AI initiative or an HR leader charged with managing change during technological disruption, this guide will equip you with the tools to create workplaces where humans and AI amplify each other's capabilities rather than compete.

Praise for
The AI-Enabled Leader

I had the privilege of being Beth's CEO early in her career and saw, even then, how her professionalism, presence, and intellect set her apart. Seeing the depth of her work today fills me with pride and admiration.

In The AI-Enabled Leader, Beth recognises and celebrates the promise of AI and the extraordinary possibilities it offers for transformation. Yet she reminds us that however rapidly technology evolves, real transformation is built on the enduring foundations of trust, human capability, and authentic leadership. Through real-world stories and insightful anecdotes, she illustrates how technology solutions too often falter when introduced without clarity, engagement, and recognition of the people at the centre of change.

As the French writer Jean-Baptiste Alphonse Karr observed, "plus ça change, plus c'est la même chose" - the more things change, the more they stay the same. Beth encourages leaders to focus on what will be essential to making AI a force for genuine and valuable transformation - the timeless basic elements of trust and authentic leadership.

Alan Singer, *IT Industry Thought Leader/Start-up Founder*

Beth is a leader in her field and bridging the human behavioural gap with technology innovation. As a Business Psychologist and consultant of 20 years, with an MBA, this book leads the way in a way no other has - the delicate balance between human behaviour, psychology, leadership and technology.

With practical, real world tools and audits, road maps and insights, it's a resource that acts as your Chief HRO, CITO, and COO - in your pocket. Carefully stepping through the mechanics, the readiness and the fundamental core of all of our jobs - humans, technology and revenue.

As a leader in Government who has People, Culture, Safety and Technology reporting to me, it's like a dictation of my brain, the strategies we're all grappling with, and the criticality of a united, informed and strategically focussed executive team on the things that matter most - adaptability, innovation, human centred design and a "we've got this" - let's follow the frameworks.

Carly Fisher, *Director – Carly Fisher Consulting*

Thankyou Beth for speaking out loud what we are all in desperate need of as leaders, consultants, Board Directors and beyond A calm and compelling reminder that even as intelligence becomes commoditised, leadership must remain human. Beth's work offers a hopeful compass for leaders navigating the tension between technological inevitability and human integrity and gives them a vocabulary for empathy and integrity at a time when it is most needed.

Tim Matthis, *Founder – AI in Australia*

Beth Barratt-Browne is a rare consultant who has stayed ahead of the rapid advances AI is bringing to every corner of business, while remaining deeply grounded in the human side of leadership. In her second book, she clearly and comprehensively outlines what leaders must consider to ensure AI and their teams thrive together. Her step-by-step framework is both practical and insightful, offering executives a thoughtful roadmap for integrating AI into their organisations. Highly recommended.

Tora Cullip, *Founder - Tora Fit*

Contents

Chapter Zero

The Moment We Are In

AI is no longer a future trend. It is a structural shift already reshaping how organisations operate, how people work, and how leaders lead. Regardless of sector or geography, every organisation now operates in an environment where technology moves at exponential speed while culture, capability and governance often lag behind.

Across the world, governments are signalling the scale of this shift. Australia's National AI Plan (2025), the EU AI Act, the United States' AI risk management frameworks, Singapore's Model AI Governance Framework and the UK's AI Safety Institute initiatives all converge on the same three themes: **capture opportunity, manage risk and build trust**.

The message to leaders is unmistakable:
AI is no longer a technology issue. AI is a leadership issue.

Leading with AI is not simply adoption. It is the new test of organisational resilience, where adaptability, clarity and disciplined judgment determine who moves forward and who falls behind. Leaders must navigate repeated moments of uncertainty, manage resistance early and decisively and maintain cultural stability while technology accelerates around them.

The organisations that succeed will not be those collecting tools. They will be those strengthening the human and organisational foundations that allow AI to be used safely, ethically and effectively. They will create clarity where others experience confusion and direction where others hesitate.

Responsible AI leadership now requires integrity, transparency and governance rigour. Cybersecurity has become inseparable from strategy. Without it, no AI innovation or adoption can remain resilient, reinforcing that ethical leadership is the foundation of sustainable success.

AI amplifies human strengths: empathy, judgment, strategic decision-making; and exposes weaknesses just as quickly. Leaders who thrive will be those who balance technological leverage with human capability, building workplaces where people and AI can contribute to performance, scale and innovation without losing what makes organisations human.

Chapter Zero introduces this global context.
The chapters that follow focus squarely on what leaders can control: building readiness, capability, trust and governance that match the moment we are living through.

This is not a book about the mechanics of AI.
It is a book about leadership in a world where technology has become a co-worker.

We are in a moment that will define organisations for the next decade.
The work of leadership begins here.

Introduction

The workplace is changing. Not gradually, not incrementally, but fundamentally.

This is not a book about technology. It is a book about leadership in a world where technology has become a co-worker.

As I sit across from CEOs and HR leaders in boardrooms around the country, I hear the same questions echoing with increasing urgency:

- *How do we harness AI's potential without losing our human edge?*
- *How do we move beyond the hype to create genuine value?*
- *And perhaps most importantly: how do we lead our people through this transformation with both compassion and clarity?*

These aren't just academic questions. They are the practical challenges keeping leaders awake at night as they navigate what may be the most significant workplace transformation since the industrial revolution.

I wrote this book because we are at an inflection point.

This moment is also a generational opportunity for HR and People & Culture leaders. For decades, the profession has been caught between operational demands and strategic aspiration. AI removes that tension. It elevates HR to where it belongs: at the intersection of people, performance, and capability. Organisations are increasingly leaning on People & Culture leaders who can blend human judgement with data intelligence, redesign roles, and workflows, and build the cultural guardrails that shape how humans and AI work together.

The organisations that thrive in the coming decade won't be those with the flashiest AI displays or the biggest technology budgets. They'll be those that most thoughtfully integrate powerful tools with uniquely human strengths.

They will treat AI not as a technology implementation, but as a **human transformation project with a technology component.**

Early in my career as a People & Culture executive, I learned a simple truth: technology amplifies culture. It doesn't fix leadership issues. It exposes them.

AI can accelerate insight, scale efficiency, and reshape work. But without trust, clarity, and capability, it can also accelerate confusion, fear, and fragmentation.

I've seen brilliant AI programs fail because leaders underestimated the emotional, identity-based, and cultural implications of change. And I've seen modest tools unlock extraordinary results because leaders intentionally built in psychological safety, redesigned roles thoughtfully, and treated people as collaborators in change, not subjects of it.

One organisation I worked with had invested millions in state-of-the-art AI customer tools. The technology worked flawlessly, technically. Yet customer satisfaction plummeted and staff turnover spiked.

Why? They launched the technology without redesigning roles, communication flows, or performance expectations. People didn't know where they fit anymore, so they pulled back. The system worked on paper, but not in the lived experience of the workplace.

With deliberate role design, leadership alignment, and trust-building, the same technology became a catalyst for improved performance and stronger human connection. That pattern repeats across industries: **AI success depends on human readiness, not capability alone.**

The rise of the enablement era

We are entering what I call the *enablement era*, a period where the true measure of leadership is not how effectively we manage people or technology, but how well we enable them to perform together.

This transformation is opening new opportunities for those in HR, leadership, and organisational development. Roles such as *AI workforce enablement, human–AI collaboration design*, and *ethical technology governance* are emerging at the intersection of people, process, and performance.

As JT Foxx and other global leadership voices have observed, the next generation of business advantage will belong to leaders who can enable human potential through technology rather than be displaced by it. The discipline of Human Resources (HR), once seen as a compliance function, now stands at the centre of this new capability frontier guiding how organisations attract, develop, and retain people in an increasingly hybrid environment.

For leaders, this marks a shift from control to enablement. The new question becomes: *How do we create the conditions for humans and AI to succeed together?*

This book is written for leaders who want to build those conditions deliberately rather than leave them to chance.

Defining what we mean by Artificial Intelligence (AI)

Before we go further, it's worth clarifying what this book means by *AI*.

Here, AI refers not to speculative futures or sentient machines, but to the practical, applied intelligence systems already reshaping work; data-driven tools that augment human decision-making, automate routine processes, and expand creative and analytical capacity.

Understanding AI in this pragmatic, people-first context helps leaders move from fear and hype to design and enablement.

Research reinforces the inference that successful AI implantation will be determined by human readiness and organisational culture resilience:

- Companies using human-centred AI approaches are **6.5x more likely to achieve positive ROI** (Harvard Business Review).
- Organisations conducting readiness assessments before deploying AI report successful outcomes **82% of the time**, compared to **23%** when they don't (Deloitte).

Despite this, many leaders still treat AI as a software rollout rather than a leadership, culture, and people capability evolution.

This book exists to help close that gap.

What matters most in this book is not the sophistication of the model, but how leaders integrate it into work, culture, and decision-making.

The shift to AI-enabled work is not just a technology movement. It is a redefinition of what it means to lead, and what it means to practice People & Culture. This era gives HR practitioners the clearest pathway they have ever had to shift from policy to strategy, from administration to value creation. Those who embrace this opportunity will shape the workforce of the next decade. Those who hesitate will find themselves reacting to change rather than guiding it.

It offers a **human-first playbook** for building trust, confidence, and capability in AI-enabled workforces rooted in real-world practice, grounded in evidence, and focused on equipping leaders at every level.

As I edit and refine this text for clarity and flow, I am mindful that several concepts within these pages: audits, governance, leading through change, and RACI frameworks among them, easily warrant a book (books!) in their own right. My aim is not to exhaust these topics, but to introduce them as essential foundations of AI-enabled leadership. At the end of the book, you will find recommended resources should you wish to explore any area more deeply or in your own organisational context. A list of further reading and resources follows the concluding chapter. And if you'd like help applying these ideas within your own organisation, I invite you to connect with me directly.

A word of caution

If your current people leadership or management practices are already under strain, pause before layering additional complexity.

AI will not fix what is broken. It will magnify it.

The fundamentals of people leadership outlined in my first book *Their Workplace, Your Business – A Founder's HR Playbook for Scalable Success* – remain the foundation for human thriving and hybrid capability. Strengthen those foundations, and AI becomes an amplifier rather than an added source of strain.

Before embarking on this journey, ensure that your people systems are clear, your leadership is accountable, and your culture supports trust and feedback. These are the preconditions for success in any hybrid, AI-enabled workplace.

How this book is structured

The journey unfolds in four parts:

Part I: The New Leadership Landscape

You'll understand the shifts reshaping work and evaluate your organisation's AI readiness, not just technically, but culturally, ethically, and psychologically.

Part II: Designing the Hybrid Workplace

You'll learn how to redesign roles, workflows, and collaboration models so humans and AI elevate each other. Not automation; *augmentation.*

Part III: Leading Humans Through Change

You'll build the leadership capability to guide people through transformation with empathy, clarity, and psychological safety; especially the middle managers who carry cultural weight.

Part IV: Scaling With Integrity

You'll explore how to scale responsibly. Here, *scaling* refers to sustaining transformation building systems, culture, and leadership capability that allow progress to multiply as distinct from scaling an organisation for sale. Measuring what matters, protecting trust, prioritising wisely, and ensuring ethics and culture stay intact as systems multiply.

Across the book you'll find:

- Real stories from organisations in transition
- Practical frameworks you can apply immediately
- Signature leadership lines to guide decision-making and conversations
- Reflection prompts to help you make meaning in your own context

How to use this book

You can read it sequentially if you're building a structured AI transformation approach, and that is your preference.

Or you can move chapter-by-chapter based on the leadership moment you're in: culture, capability, governance, resistance, measurement, or scaling.

This is not a manifesto. It is a **leadership field guide** for the real world.

You don't need to be an AI expert.

You need to be a leader who can:

- Hold trust amidst uncertainty
- Build clarity before capability
- Support identity and confidence as roles evolve
- Balance speed with stewardship
- Put humanity at the centre of technology shifts

AI won't replace leaders.

But it will **reveal** which ones understand that technology is a tool; and trust is the true infrastructure of growth.

The future belongs to leaders who treat AI as a partner and humanity as the purpose. This journey isn't about choosing between *human and machine*.

It's about **designing workplaces where both can rise**.

In 2025, the Australian Government released its first National AI Plan, reinforcing the importance of responsible adoption, worker safety, transparent governance and trust-centred transformation. These foundations mirror the leadership principles explored throughout this book.

Part I | THE NEW LEADERSHIP LANDSCAPE

Chapter 1:

The New Work Equation

Humans & AI in Modern Work

Work is changing fast. Faster than most leaders expected.

The shift we are in is not just technical. It is behavioural, cultural, structural, and deeply human.

Across boardrooms, executive offsites and leadership coaching conversations, the same questions keep surfacing:

- How do we use AI well without losing what makes us human?
- How do we stay competitive without overwhelming our people?
- How do we build capability and confidence, not fear and fatigue?
- How do we increase our efficiency without sending work offshore?

This is not a hype cycle. It is a shift in how value is created, how decisions are made, and how people work and learn. It is also an extension of capability management.

A human-centred approach means designing systems, workflows, and decisions that protect human judgment, wellbeing, capability, and accountability as technology evolves.

Leaders are not preparing for a future workplace.

We are already inside it.

The Real Shift

Previous industrial revolutions replaced physical labour.

This one touches thinking work, creative work, and decision-making.

AI can now ideate, analyse, recommend, and optimise.

And yet, there are clear limits.

AI struggles with nuance, context, ethics, ambiguity, and care. Like people, AI also can get overloaded and will sometimes make things up when information is missing.

AI is also teaching humans how to communicate clearly. If you cannot describe what you want, AI cannot produce it. Prompting exposes unclear thinking, hidden assumptions, and unspoken expectations. Used well, AI strengthens the very human skills leaders need more of.

So, the leadership question is not, “How do we automate?”

It is, “How do we design work where humans and AI lift each other to create outcomes neither could achieve on their own?”

That is the new work equation.

A Leadership Inflection Point

Research from Stanford, MIT and Harvard is consistent. The organisations getting this right treat AI as a human transformation supported by technology, not a technology project with humans on the side.

Leaders who succeed in this phase:

- Protect psychological safety before introducing tools
- Build clarity and capability together
- Strengthen judgment, not bypass it
- Engage middle managers early, not as an afterthought
- Treat AI as a partner, not a replacement

Australia's National AI Plan echoes this shift by framing AI as a whole-of-economy capability where leadership, not technology, determines success. The future of work will be shaped as much by human judgment as by machine intelligence.

AI does not reduce the importance of leadership.
It increases it.

Understanding the Human and AI Relationship

We need language for how humans and AI work together.
This is not a binary choice between "AI replaces jobs" or "AI assists humans", but a spectrum of collaboration models.

A simple way to frame this:

Human-led

Humans decide. AI gives insight or shortcuts.
Examples: legal research, drafting, audit preparation.

AI-assisted

AI suggests or ranks options. Humans choose.
Examples: customer service prompts, risk flags, talent screening.

True partnership
Humans and AI iteratively work together.
Examples: design exploration, forecasting, scenario testing.

AI-led with human oversight
AI executes. Humans supervise and intervene if needed.
Examples: automated approvals, supply chain optimisation.

Human-in-the-loop means AI can propose or execute steps, while humans retain review authority for higher-risk decisions and remain accountable for outcomes.

Autonomous systems
Only in low-risk, well-controlled environments.
Humans remain accountable.

Every effective hybrid workflow relies on trust: trust in technology, trust in humans supervising it, and trust that leaders have designed the right level of collaboration for the risk involved.

The most effective organisations do not ask, "Can we use AI here?" They ask, "What level of collaboration makes sense for the risk, purpose and context?"

A Quick Story from the Field

A financial services team introduced AI to triage client inquiries.
The tech worked. The experience did not.

Customers felt pushed away.
Staff felt like switchboard operators instead of advisors.

The organisation paused and redesigned:

- AI handled first-pass sorting
- Humans engaged earlier and with better context
- Advisors trained to use AI insights live in the conversation

Customer satisfaction lifted.
Staff confidence rose.
The AI became a co-pilot rather than a barrier.

The lesson:
AI needs workflow design, not just deployment.

What the Research Tells Us

Industry data across leading firms is becoming very clear:

- Successful AI adoption is **six** times more likely in human-centred organisations
- Over 70 percent of failed AI programs stem from culture, clarity and leadership gaps, not technical faults
- Organisations that **define decision roles** between AI and humans outperform those that "experiment and hope"
- Trust and capability predict AI success more than budget or tooling

In short:

AI maturity follows leadership maturity.

New Leadership Requirements

To succeed, leaders build:

- Clarity before scale
- Confidence before automation
- Guardrails before acceleration
- Learning culture before tooling culture

And they answer three questions early:

1. What remains uniquely human here?
2. Where can AI safely accelerate?
3. How do we maintain trust and accountability?

Signature Leadership Lines

- AI reveals leadership maturity, it does not replace it
- Clarity fuels confidence
- You do not adopt AI, you *redesign* work with it

Reflection

What assumptions about AI are shaping our current decisions, and which of them need to be challenged to support a more human-centred approach?

Where does your team need clarity and reassurance before you introduce (more) AI into workflows?

Looking Ahead

In the next chapter, we look at readiness.

Not as a technology checklist, but as a leadership responsibility.

Readiness is not about enthusiasm.

It is about alignment, trust, and capability.

If Chapter 1 gives us the mindset, Chapter 2 gives us the map.

Chapter 2:

Are You Ready?

The Leadership & Workforce AI Readiness Audit

Most AI transformation failures are not technology failures. They are readiness failures.

Many organisations rush toward AI because the opportunity feels urgent. The market moves, competitors adopt tools, headlines accelerate. Pressure builds. Leaders move fast, believing speed equals progress.

But speed without readiness often leads to expensive lessons.

Take, for example, a financial services organisation that deployed an AI customer support system with enormous excitement. They had budget, vendor support, executive sponsorship. On paper, they did everything right.

Six months later, the system was offline, customer complaints had spiked, and employees felt replaced, not supported. The technology worked. The environment around it did not.

Their mistake was not ambition. It was sequencing. They focused on installing the tool before preparing the organisation. They invested in technology before investing in people and readiness.

This chapter aims to help you avoid that trap.

Readiness is not technocratic bureaucracy. It is leadership discipline. It is how you ensure people, structure, culture, and systems are prepared to adapt, not resist. It is how you build confidence, capability, and trust before asking your teams to work differently.

Accelerating before preparing people does not speed adoption. It slows it.

Why Readiness Matters

AI implementation succeeds or fails long before tools are purchased or workflows are redesigned. The foundation is readiness: cultural, psychological, technical and ethical. As research from Stanford's Digital Economy Lab consistently shows, successful AI implementations begin with structured readiness assessments that evaluate not only technical capacity but human and cultural conditions.

Like capability development in any domain, readiness will be strengthened through assessment, practice, and reinforcement. As your organisation evolves, reassessment ensures alignment, confidence and resilience. People change, processes change, and technology changes. A periodic pulse check keeps your AI journey grounded in real conditions, not assumptions.

Research consistently reinforces this truth:

- Organisations that conduct structured AI readiness assessments are more than three times more likely to achieve successful outcomes (Deloitte, 2023).

- Stanford's Digital Economy Lab reports that human-centred readiness is the strongest predictor of sustainable AI value, outranking data maturity, and budget.
- McKinsey finds that the leading cause of failed AI programs is not the algorithm or model quality. It is unclear leadership ownership, low trust, and lack of workforce preparation.

In simple terms: readiness is return-on-investment insurance.

It protects your people, your strategy, and your credibility.

What We Are Really Assessing

Readiness is not about perfection. It is about clarity.

It asks three questions:

1. Are our leaders aligned, informed and accountable?
2. Is our culture prepared to learn, adapt, and trust?
3. Do our people have the confidence and capability to work with AI?

If you cannot answer yes with confidence, the opportunity is not to pause. It is to prepare.

The Readiness Audit

Below are three practical tools to evaluate your organisation. You can complete them privately, with your executive team, or through a facilitated process.

Honest assessment enables strategic sequencing.

Leaders do not need to know everything. They need to know where to start.

Leadership Readiness Index

Purpose: Assess decision clarity, accountability, and ethical leadership foundations.

Score each item: Yes / Somewhat / Not yet

Leadership Capability	Self-Check Statements
Strategic clarity	We have a shared view on why we are adopting AI and where it creates value
Decision rights	We know who approves AI use cases and who remains accountable
Ethical guardrails	We have principles for when and how we want AI to support decisions
Oversight	Leaders understand where human judgment must prevail
Communication	Executives can explain AI decisions confidently and simply
Risk posture	We know our appetite for experimentation versus caution

Cultural Readiness Assessment

Purpose: Understand whether trust, safety and transparency exist to support change.

Score each item: Yes / Somewhat / Not yet

Cultural Factor	Self-Check Statements
Trust and transparency	Employees believe leaders will use AI fairly and responsibly
Sense of safety	People feel safe asking questions and expressing concerns
Purpose alignment	People understand why AI is being adopted and what it means for them

Cultural Factor	Self-Check Statements
Change resilience	Our culture supports learning, feedback, and iteration
Data confidence	Employees trust how data is used and protected
Role security	Staff see AI as support, not replacement, when communicated effectively

Capability and Confidence Scan

Purpose: Identify skills for hybrid human-AI work.

Score each item: Yes / Somewhat / Not yet

Capability	Self-Check Statements
AI literacy	Teams understand what AI can and cannot do
Data literacy	People can interpret, question and work with system insights
Human judgment	People know when to override AI and how to explain their decisions
Workflow fluency	Teams know how AI integrates into their tasks and processes
Learning readiness	People are willing to explore, experiment and upskill

How to Interpret Your Results

Do not treat this as a pass-fail test. It is a compass.

Score Mix	Meaning	Action Approach
Mostly Yes	Strong foundation	Begin piloting with governance and communication
Mix of Yes and Somewhat	Emerging readiness	Build clarity and education alongside pilots

Score Mix	Meaning	Action Approach
Mostly Somewhat	Developing	Prepare systems, people, and communication before scaling
Mix of Somewhat and Not Yet	Early-stage	Run small learning pilots, start with literacy and trust
Mostly Not Yet	Foundational stage	Begin with leadership alignment, purpose narrative, and safe learning environments

Readiness is not about slowing down. It is about sequencing intelligently.

Start where the organisation is. Build confidence before scale. Provide support before expectation.

Building Your Readiness Plan

Once you assess your organisation, define:

1. **Where we can start now**
   Pilot areas where readiness is strongest.
2. **Where we must prepare first**
   Invest in clarity, communication, and confidence.
3. **How we build momentum**
   Quick wins early, learning loops always.
4. **How we support people**
   Change happens at the pace of trust, not technology.

Key Insights for Your Readiness Journey

- **Readiness is cultural as much as technical.** Trust, alignment and psychological safety matter as much as tooling.
- **Clarity reduces anxiety.** When people understand why AI is being introduced and what remains uniquely human, confidence increases.
- **Capabilities grow through practice.** Start small, learn, adapt.

- **Leaders set the conditions**. Sponsorship, messaging and accountability determine momentum.
- **Readiness changes over time**. Reassess at key transition points as systems and roles evolve.

Signature Leadership Lines

- You can't automate rediness
- Clarity beats complexity
- Culture is the first system you design

Reflection

Where does my organisation need clarity before tools?

Which group requires support first: leaders, middle managers, or teams?

How will I communicate not just what AI is, but what it means for people?

Where will I pilot, and where will I prepare?

Write your answers. They become your roadmap.

Closing Thought

AI readiness is not a technical milestone.
It is a leadership discipline.

Readiness gives you your starting point. The next step is designing the team structures where humans and AI elevate each other's strengths.

Technology moves fast. People move on trust.

Prepare your organisation thoughtfully and you will accelerate faster than those who rush without readiness.

True transformation is earned, not installed. **Readiness is how you earn it**.

Part II | DESIGNING THE HYBRID WORKPLACE

Chapter 3:

Designing Hybrid Teams

Roles, Workflows & Collaboration

The most successful hybrid workplaces are not defined by the technology they use. They are defined by the clarity with which leaders design how humans and AI work together.

This is not about replacing jobs with digital tools. It is about redesigning work so that human strengths and machine capabilities amplify each other.

Future-focused organisations are not asking "Where can we automate?" They are asking "How do we structure teams, so people and technology elevate performance together?"

Australia's policy direction reinforces the need for clear human accountability. As AI systems take on more operational load, hybrid roles must be deliberately designed so that judgment, context and ethical reasoning remain human strengths, not casualties of automation.

Why this matters

Designing hybrid teams matters because the way we assign work between humans and AI determines whether technology becomes an accelerator of capability or a source of confusion and mistrust. When roles and workflows are unclear, workload inequity grows and psychological safety weakens. When they are intentional, clarity, capability, and trust increase together.

Research from Harvard Business Review has shown that organisations that intentionally design for human and AI collaboration achieve significantly higher productivity and engagement than those that simply deploy automation tools. McKinsey has also reported that the most successful AI transformations are those that redesign roles, workflows, and decision-making models rather than trying to bolt AI onto legacy processes.

This chapter moves us from readiness to action. If Chapter 2 helped you understand where you are, Chapter 3 helps you architect what comes next.

A leadership perspective

When I begin hybrid design work with executive teams, the first question I often ask is simple:
"*Who decides what work belongs to a human and what belongs to AI?*"

Most rooms fall silent.

Not because leaders lack capability.
But because this work is new, nuanced, and requires strategic clarity.

Without intentional design, AI adoption risks creating confusion, inequality in workload, and a loss of trust. With thoughtful structure, it creates capacity, clarity, and energy for more human work.

A story from the field

A professional services firm I worked with had rolled out an AI knowledge assistant. Technically brilliant, it surfaced case studies and expert content instantly. Yet adoption stalled. Workloads did not shift. Feeling threatened, experts withheld knowledge from the system.

Instead of pushing harder, we paused and redesigned:

- We defined the role of a **Knowledge Curator**, responsible for refining and validating inputs
- We introduced a **Client Insight Partner** role, where consultants focused on strategy and relationship work supported by AI research
- We created a clear workflow handoff: AI collected, humans analysed and advised
- Experts became mentors to the system, not competitors to it.

Within 90 days, usage doubled, client satisfaction improved, and staff reported feeling more valued, not less.

Success came not from more technology, but from **redefining roles, responsibilities, and purpose**.

The Human-AI Capability Spectrum

To design hybrid teams well, leaders must start with capability, not tasks.

Humans excel at

- Judgment and context
- Relationship building and trust

- Creativity and sense-making
- Ethical reasoning and values-based decisions

AI excels at

- Pattern recognition, analysis, and speed
- Consistent execution
- Data processing and forecasting
- Reducing cognitive load

Hybrid performance happens when both are intentionally combined.

This shift moves us from job descriptions to **capability architecture.**

Emerging Role Archetypes for Hybrid Workforces

Drawing on Stanford's Human-Centred AI research, five useful role patterns are emerging:

Archetype	Value
Trainer	Improves AI quality through human feedback
Explainer	Interprets AI output for stakeholders
Sustainer	Maintains ethics, quality, and alignment
Complementor	Applies human strengths supported by AI
Overseer	Maintains accountability for outcomes

Most roles will blend elements of these. The key is clarity, not volume.

Workflow redesign

Key design questions

1. Which decisions belong to AI, which belong to humans, and which are shared?
2. What information must move between AI and people to maintain quality?

3. What triggers handoffs or human review?
4. How do we build learning loops into the workflow?

Hybrid workflow principles

- Let AI handle the predictable
- Let humans lead the ambiguous
- Create structured handoffs and escalation pathways
- Track improvements, overrides, and exceptions as learning fuel

The Human in the Loop: Keeping Judgment at the Core

As AI systems take on more operational and analytical work, maintaining a *human in the loop* ensures that ethical reasoning, contextual judgement, and empathy remain active in every critical decision. This is not about slowing progress; it is about protecting trust and quality.

A human-in-the-loop model includes three elements:

- **Oversight**: Humans review and approve AI-generated outputs in areas with significant impact such as hiring, performance evaluation and customer decisions.
- **Intervention**: People have clear authority to question or override AI recommendations when context, fairness, or values are at stake.
- **Learning loops**: Human feedback is used to refine algorithms, improving both system performance and workforce capability over time.

This approach is increasingly recognised as a cornerstone of responsible AI use. Research from *Stanford's Human-Centred AI Institute* shows that systems with structured human oversight consistently achieves higher accuracy and greater user trust than fully automated models.

The most effective hybrid teams treat AI as an intelligent collaborator, not an autonomous authority.

Governance that protects trust and accelerates progress

Hybrid teams thrive when **accountability stays human** and AI participation is transparent.

Effective governance includes:

- Clear ownership of system behaviour and outcomes
- Transparency about how decisions are influenced by AI
- Oversight committees balancing ethics, risk, and performance
- Regular calibration reviews

This is not bureaucracy. This is confidence, safety, and momentum.

Putting it all together: practical steps

To design a hybrid team, leaders need to:

1. Map key work to human, AI, or shared responsibilities
2. Define new or adapted role profiles
3. Document decision rights and escalation triggers
4. Establish handoff protocols and feedback loops
5. Create transparent accountability and review structures
6. Measure value, trust, adoption, and capability growth
7. Start with one workflow. Learn. Adapt. Expand.

Signature Leadership Lines

- Hybrid success is designed, not discovered
- Structure creates trust
- Technology serves meaning, not the other way around

Reflection

Where are roles or workflows unclear today, creating friction or hesitation?

Which decisions genuinely belong to AI, and which must remain human-led?

Where could a simple handoff, escalation point, or learning loop strengthen trust?

What is one workflow we could redesign as a starting point for hybrid collaboration?

How will I redesign work, so humans thrive and technology elevates, not replaces?

Closing thoughts

Hybrid workplaces do not emerge organically. They are designed with intention, clarity, and respect for the strengths of both humans and AI. When leaders define roles, decision rights, and collaboration pathways, trust grows and capability expands.

The highest-performing hybrid organisations do not chase technology. They architect human potential and let technology support it.

AI will accelerate execution. People will always anchor meaning, trust, and progress.

This is not a tech chapter. It is a leadership chapter.

The question for you now is not "*Which tools?*"
It is "*How will I redesign work, so humans thrive and technology elevates, not replaces?*"

That is hybrid leadership.

In the next chapter, we step deeper into governance, ethics, and safety; the guardrails that make hybrid work sustainable, fair, and trusted. Clear roles are the beginning. Clear guardrails are the safety net.

Chapter 4:

Governance, Safety & Fair Work in the AI Era

The integration of AI into workplace systems has created a governance challenge unlike any before it, one that stretches beyond compliance frameworks and into the heart of organisational trust.

This is not simply about technology risk. It is about who holds responsibility when systems learn, adapt, and make consequential decisions affecting people's livelihoods. As AI becomes embedded across workflows, the question has shifted from whether we use it to **how we govern it responsibly**.

Why this matters

AI differs fundamentally from traditional workplace technologies. Unlike static systems that follow programmed rules, modern AI learns from data, adapts, and acts with varying degrees of autonomy. That makes it both powerful and unpredictable.

Traditional governance approaches, built for predictable systems, are not suited to tools that evolve.

AI governance is how values travel through systems. Without it, technology accelerates risk faster than it accelerates performance.

Research from *Harvard Business Review* shows that organisations with intentional AI governance achieve higher trust and engagement. *McKinsey* found that fewer than one in four organisations have comprehensive governance frameworks in place, even though those that do are more than twice as likely to realise significant business value from AI.

Without appropriate guardrails, AI systems can amplify bias, make consequential decisions without transparency, and erode trust among employees who feel managed by algorithms rather than supported by leadership.

Australia's National AI Plan, launched in December 2025 reinforces this emphasis on governance by highlighting transparency, accountability and documentation as essential safeguards for responsible AI use. As organisations adopt more complex systems, leaders are expected to make governance visible, auditable and aligned with clear human accountability. Strong governance is no longer a technical preference. It is a leadership requirement.

A story from the field

A large Australian financial services firm implemented an AI-powered performance system to identify high-potential employees. Within months, the algorithm began producing recommendations that penalised staff returning from parental leave. It was not designed to discriminate; it simply learned from historical data shaped by old inequities.

Once an AI governance framework was established, bias detection became part of the review process. Promotion and pay decisions stabilised, and trust was restored. The lesson was clear: effective governance does not restrict innovation; it protects fairness and integrity while enabling confident adoption.

Ethical Governance Frameworks:
Building Accountability in Algorithmic Systems

Organisations that establish governance frameworks early experience fewer ethical and performance issues than those that retrofit controls later. Governance is not an IT function; it is a leadership responsibility that shapes how fairness, accountability, and human judgment show up in every AI-enabled decision.

Effective frameworks rest on three pillars: **structural accountability, procedural transparency, and outcome responsibility**.

1. **Structural Accountability**
   Structural accountability defines who owns responsibility for AI systems throughout their lifecycle, from design through deployment and review.
   At the executive level, this typically means creating an **AI Governance or Ethics Committee** with cross-functional representation from technology, legal, HR, and business areas.
   Operationally, organisations benefit from implementing a **RACI-E matrix** (Responsible, Accountable, Consulted, Informed, Escalation) to ensure clear ownership and rapid escalation when systems behave unexpectedly.
   One financial services organisation introduced this model after its fraud detection AI produced disproportionate results across regions. Once accountability roles were defined, similar issues were resolved within days rather than months.

2. **Procedural Transparency**

   Transparency ensures that decisions made or informed by AI can be explained and understood.

   This does not require exposing complex code; it means providing clarity appropriate to each audience. Technical teams need documentation of data sources and models. Leaders need summaries of key factors influencing decisions. Employees need accessible explanations when AI affects them directly.

   An Australian healthcare organisation adopted a tiered transparency model: detailed data for clinicians and plain-language summaries for patients. Each group received clarity suited to their role, strengthening both trust and engagement.

3. **Outcome Responsibility**

   Governance must also include mechanisms for continuous monitoring and remediation. Regular bias audits, error tracking, and post-implementation reviews ensure accountability does not end at deployment. Responsible organisations use these insights to improve both systems and human processes, keeping fairness and quality at the centre.

Safety Protocols and Risk Mitigation

AI introduces new forms of workplace safety, encompassing technical, operational, informational, and psychological domains. Comprehensive safety frameworks protect both organisational integrity and employee wellbeing.

1. **Technical Safety**

   Technical safety safeguards reliability, security, and performance stability. This includes ongoing monitoring for model drift, cybersecurity protections, and fail-safe mechanisms to maintain core functionality during unexpected behaviour.

2. **Operational Safety**
   Operational safety focuses on how AI interacts with human workflows. Effective systems include documented handoffs, override controls, and escalation pathways for exceptions. Organisations that define these boundaries early experience fewer disruptions and greater confidence among staff.

3. **Informational Safety**
   Informational safety ensures data privacy, governance, and appropriate transparency about how data shapes AI outputs. Employees and customers alike need to understand what information is being used, why, and how it influences outcomes.

4. **Psychological Safety**
   Beyond systems and data, psychological safety underpins sustainable AI adoption. Leaders must communicate openly about why and how AI is being introduced, provide space for questions and feedback, and support employees as they adjust to new ways of working. Confidence and clarity are protective factors in change.

Fair Work Principles in the AI Era

As AI begins to influence decisions about performance, scheduling, allocation and opportunity, fairness can no longer be assumed. It must be deliberately designed.

Fair work is not an outcome of good technology; it is the result of deliberate leadership choices. As automation evolves, fairness must evolve with it.

Drawing from the *Fair work* framework and field experience, five principles underpin fair work in AI-enabled organisations:

1. **Fair Compensation**: Recognise hybrid capability and share productivity gains equitably.

2. **Fair Conditions**: Ensure safe, healthy, and psychologically secure work environments.
3. **Fair Contracts**: Maintain clarity and transparency in algorithmic or platform-based employment.
4. **Fair Management**: Preserve access to human review and appeal in decisions made or influenced by AI.
5. **Fair Representation**: Involve employees in shaping the systems that affect their work.

Organisations that adopt these principles report stronger trust, engagement, and technology uptake. **Fairness is not a cost; it is a multiplier of performance.**

Signature Leadership Lines

- Governance is culture codified
- Transparency scales trust
- Accountability must stay human

Reflection

Where in our organisation is AI influencing decisions today without enough clarity or visibility?

Which governance responsibilities in our organisation are clear, and which need definition?

Do our people know when and how they can question or challenge an AI-supported outcome?

What simple rhythm (monthly review, quarterly bias audit, decision log) would improve trust immediately?

Leadership Closing Thoughts

Governance is the quiet infrastructure that makes innovation sustainable. It turns uncertainty into confidence and ensures that technology strengthens rather than substitutes human judgment.

As AI becomes a co-worker across more of the organisation, leadership must stay visible, values-led, and accountable. In the next chapter, we move into the human capability side of hybrid work, exploring how confidence, learning and psychological safety turn AI from disruption into growth.

Governance creates the boundaries. Capability gives people the confidence to operate within them.

Chapter 5:

Capability, Confidence & Human Potential

The greatest competitive advantage in the AI era will not come from technological sophistication alone. It will come from organisations that deliberately build the human capabilities that machines cannot replicate.

While much attention has focused on technical readiness, forward-thinking leaders recognise that the future belongs to those who systematically develop capability, confidence, and psychological safety; creating the conditions where humans and AI elevate one another. The answer lies not in treating AI implementation and human development as separate initiatives, but in **recognising their profound interdependence**.

Australia's emerging national AI strategy places equal weight on human capability and technical investment. Skills, confidence and adaptive judgment remain the primary differentiators of performance as AI tools become more widespread.

This chapter explores the intersection between technological advancement and human development. It examines how leaders can strengthen the skills, confidence, and mindset that allow people to thrive alongside AI rather than compete with it.

Why this matters

Research from *Harvard Business Review* shows that organisations achieving the greatest returns from AI investments are those that invest equally in human capability development. *Deloitte* found that companies with structured capability programs realise 34% higher returns from their AI initiatives than those focused solely on technology.

Technology enables potential, but human confidence unlocks it. Without environments that foster psychological safety and continuous learning, even the best AI systems will fail to deliver sustainable performance.

Beth Comstock, former Vice Chair of GE, argues, the real waste of emerging technology is using it only to make old processes faster or cheaper instead of unlocking what was previously impossible.

The question for leaders is no longer *what* AI can do, but *how* humans can grow with it.

A story from the field

When a major Australian bank introduced AI-powered customer service tools, initial efficiency gains were impressive. And short-lived. As staff began to rely on automation, contextual understanding and empathy began to decline, eroding customer satisfaction.

Leadership recognised that technical implementation was not enough. They pivoted to develop their people's capabilities in emotional intelligence, ethical judgment, and contextual reasoning alongside the AI rollout. Within months, performance rebounded and surpassed expectations.

The result was a hybrid system where AI handled data and routine interactions, freeing humans to focus on empathy, complex resolution, and relationship depth. Neither could have succeeded alone.

(This example is based on composite insights drawn from publicly available case reports and observed industry practice rather than a single public case study.)

Developing Future-Proof Human Capabilities

Organisations that thrive in the AI era do not focus on what machines can do, but on what only humans can do well.

Drawing on *MIT* and *World Economic Forum* research, the **Complementary Skills Framework** identifies three key dimensions for prioritising human capability development:

1. **Uniquely Human**: Skills that AI cannot replicate effectively.
2. **High Value**: Skills that deliver measurable business impact.
3. **Developable**: Skills that can be systematically cultivated through learning and practice.

Leaders can map work activities using four quadrants to guide development focus:

- **Automate** – Routine, rule-based tasks where AI excels.
- **Augment** – Analytical or interpretive work where humans and AI collaborate.
- **Enhance** – Creative, strategic, or relational work strengthened by AI insights.
- **Preserve** – Deeply human domains such as trust, ethics, and meaning-making.

Most organisations find their highest value opportunities in the *Augment* and *Enhance* quadrants, where collaboration creates outcomes neither side could achieve independently.

The Core Human Capability Domains

Research consistently identifies several capability domains that will remain critical and increasingly valuable in an AI-augmented workplace:

1. **Complex Problem Solving**

   AI can analyse data at scale, but humans frame the right questions. Effective problem solvers interpret ambiguity, weigh competing values, and connect insights across disciplines. This capability grows through exposure to diverse challenges and reflection on judgment-based decisions.

2. **Social Intelligence**

   Empathy, negotiation, and collaboration remain uniquely human strengths. AI may simulate tone, but it cannot read intent, emotion, or nuance the way humans can. These capabilities are foundational to trust, leadership, and culture cohesion.

3. **Creativity and Adaptability**

   Where AI optimises existing processes, humans imagine new ones. Encouraging experimentation, curiosity, and the ability to learn from failure builds the creative confidence needed for continuous reinvention.

4. **Ethical Reasoning**

   As AI influences more decisions, ethical reasoning becomes a core business skill. Employees must be equipped to question not just whether something *can* be done, but whether it *should* be.

Psychological Safety and Confidence

The introduction of AI into daily work brings both excitement and unease. For many, uncertainty about relevance, capability, and control becomes a barrier to adoption.

Psychological safety, the belief that one can ask questions, express doubt, or make mistakes without fear of judgment, is the foundation that turns anxiety into learning.

Professor Amy Edmondson's research shows that teams with high psychological safety outperform others on innovation, adaptability, and retention. This holds especially true in AI environments where experimentation and iteration are constant.

The most successful organisations address the **confidence gap**. The distance between what AI can do and how confident people feel working alongside it. Leaders bridge this through five practices:

1. **Transparent Purpose** – Explain not just what AI will do, but why it matters and how it supports human work.
2. **Co-Design and Involvement** – Involve employees in shaping AI workflows to create ownership.
3. **Learning and Experimentation** – Encourage safe trial and error, treating mistakes as insights.
4. **Recognition and Reinforcement** – Celebrate examples where human judgment improves AI outcomes.
5. **Open Dialogue** – Create channels for feedback, questions, and continuous conversation about change.

Confidence is not taught in a workshop; it is built through trust and participation.

Reimagining Career Development in Hybrid Workplaces

Traditional career ladders are giving way to **skill pathways** that reflect fluid, multidimensional work.

The **Skill Adjacency Model** helps organisations visualise how people can grow laterally across emerging domains rather than narrowly up a hierarchy.

By mapping adjacent skills – capabilities that build on existing strengths and open new directions – employees can move fluidly between roles as technology evolves.

A major Australian bank applied this model during its AI transition. Instead of restructuring jobs, it identified adjacent skills in data literacy, ethical decision-making, and relationship management. Employees were able to move toward either analytical or relational specialisations, maintaining engagement while evolving capability.

Research from *Deloitte Human Capital* confirms that organisations using skill-based development models achieve 55% higher retention during transformation. Career development becomes less about promotion and more about adaptability.

Building Learning Ecosystems

In hybrid workplaces, learning must become continuous, embedded, and responsive.

Effective learning ecosystems integrate:

- **Microlearning**: Short, accessible modules at the point of need.
- **AI-Powered Coaching**: Personalised development feedback based on performance patterns.

- **Collaborative Learning Communities**: Peer-based networks that connect experience across functions.
- **Experiential Learning**: Real-world practice linked to live challenges.
- **Reflection**: Guided review to convert action into insight.

As *McKinsey* notes, organisations that integrate learning into the flow of work build capability four times faster than those relying on formal training alone.

This is where capability and confidence reinforce each other – a true hybrid system.

Signature Leadership Lines

- Confidence is the new capability
- Skill without confidence stalls progress
- Leaders turn uncertainty into learning

Reflection

Where is my workforce currently strongest across capability, confidence, and adaptability?

Where are fear, uncertainty, or low confidence slowing adoption?

Which capability domains (problem solving, social intelligence, creativity, ethics) matter most for our context?

What is one practical step we can take this quarter to strengthen hybrid human–AI performance?

Leadership Closing Thoughts

Sustainable performance in the AI era depends on more than digital skill. It depends on confidence, adaptability, and a culture where humans are empowered to think, learn, and lead.

The evidence is clear: organisations that develop human capability alongside technology achieve higher returns, stronger trust, and greater resilience.

Capability building without confidence creates compliance, not transformation. Confidence without learning creates optimism without progress. Together, they form the foundation for hybrid success.

When organisations invest in both, they unlock the uniquely human strengths that make hybrid workplaces thrive. In the next chapter, we turn to the leadership behaviours required to guide people through uncertainty, resistance, and change — the emotional landscape that determines whether AI becomes a source of fear or a source of momentum.

Capability enables performance, but trust enables change.

Part III | LEADING HUMANS THROUGH CHANGE

Chapter 6:

Leading Through Change

Trust, Resistance & Psychological Safety

The technical work of AI is often simpler than the human work. Fear, uncertainty, and resistance surface quickly when roles, status and identity feel at risk. Change in an AI-enabled workplace is rarely about technology. It is about identity, confidence, and a sense of place in the future.

Adoption depends less on algorithmic sophistication and more on a leader's ability to build trust, meet resistance productively, and create environments where people can learn without fear.

Middle managers carry the heaviest load. They are asked to translate strategy, keep performance steady, reassure anxious teams, and master new systems at the same time. When equipped, they become the catalysts of change. When neglected, progress stalls.

Why this matters

- Organisations that build trust before rollout see faster adoption and higher satisfaction than those that "explain the tech and hope."
- Resistance is rarely technophobia. It is often information about risks, workflow realities, and identity concerns. Treat it as a data source.
- When people hesitate, they are not resisting the tool. They are protecting their status, competence, and relevance. Read resistance as insight, not obstruction.
- Psychological safety converts uncertainty into learning. Without it, teams perform compliance, not change.

Trust is the pre-condition for transformation. Resistance is data, not defiance. Psychological safety converts uncertainty into learning and ensures that people perform change, not compliance.

The Trust Equation for AI transitions

Use this practical lens to diagnose and strengthen trust:

Trust = (Credibility × Reliability × Intimacy) ÷ Self-Orientation

- **Credibility**: Do leaders and sponsors demonstrate real understanding of both capabilities and limits? Are explanations plain-language and honest?
- **Reliability**: Do actions match promises on timelines, training, and success measures? Are guardrails and review points visible?
- **Intimacy**: Have you created spaces where questions and doubts can be voiced without penalty?
- **Self-Orientation**: Is the narrative only about efficiency and cost, or also about dignity, growth, and meaningful work?

Leaders often over-invest in credibility and under-invest in intimacy. Balance both.

Signals you're getting it right: teammates can articulate the "why," leaders acknowledge limits, feedback loops exist, and there is a clear path to contest or review AI-influenced decisions.

Middle managers as change catalysts

Middle managers sit at the fulcrum. Equip them across three dimensions.

1. **Knowledge (beyond AI 101)**
   - Function-specific use cases and limits
   - The implementation roadmap and realistic impacts on workflow, metrics, and roles
   - "Tiered decision rights" so they know what remains human, what is shared, and what is automated

2. **Skills for hybrid leadership**
   - Coaching teams to balance model output with context and judgment
   - Facilitating human-AI collaboration rituals: daily syncs, exception reviews, post-override learnings
   - Narrative craft: explaining why the change matters in their team's language

3. **Mindset shift: from control to orchestration**
   - Less "I decide everything," more "I design the system where good decisions happen"
   - Confidence to pause a rollout when trust debt appears, then restart with fixes

Manager Experience Lab (template):

- Hands-on sandbox with the real tool and synthetic cases
- "What could go wrong here?" challenge session
- Draft team narrative and Q&A sheet
- Pre-mortem: "If this failed in 90 days, why?"
- Commitments: what managers will say, show, and measure in week one

Resistance as a resource

Reframe pushback as signal, not sabotage.

The Resistance Intelligence Framework

1. **Seek opposition early**: anonymous channel, challenge labs, pre-mortems.
2. **Mine patterns**: cluster concerns by process, role, metric, or equity risk.
3. **Respond visibly**: publish "You said, we did," adjust plans, close the loop.

Typical roots and matching responses:

- **Identity and value**: clarify what becomes more human, not less; show new craft standards.
- **Autonomy and control**: define override paths and meaningful human review.
- **Practical viability**: run pilots in messy realities, not lab conditions, and share results.

Australia's National AI Plan reinforces this reality. Middle leaders are now recognised as pivotal to adoption, trust and workplace safety. Equipping them is not optional. It is the leverage point for meaningful, human-first transformation.

Psychological safety in practice

Psychological safety remains the strongest predictor of whether people engage with AI, learn from it and feel empowered to experiment without fear of judgment.

Make learning safer than silence.

- **Leader vulnerability**: name what is unknown, invite correction, model asking for help.

- **Learning zones**: time-boxed experiments with explicit permission to try, fail, adjust.
- **Rituals that stick**: daily ten-minute "exceptions and insights," weekly "overrides that improved outcomes," monthly "what we've changed."
- **Fairness by design**: transparent criteria for any AI-influenced assessment, appeal routes, and privacy boundaries.
- **Confidence builder**: celebrate moments where human judgment corrected or contextualised AI outputs and improved results. Tell those stories often.

These practices form the human foundation of the enablement era, where clarity, trust and learning matter as much as any technical milestone.

Two-week starter plan

Week 1

- Announce purpose, guardrails, and decision rights.
- Run Manager Experience Lab.
- Stand up feedback and contestability channels.
- Begin baseline measures: adoption, exception rate, trust pulse.

Week 2

- Pilot in one workflow with clear entry/exit criteria.
- Hold daily exception huddles and log learnings.
- Publish "You said, we did" updates.
- Decide to scale, iterate, or pause based on evidence and trust signals.

Measures that matter

Track both performance and people:

- **Performance**: processing time, accuracy, exception quality, override outcomes

- **People**: psychological safety pulse, perceived fairness, manager confidence, enablement scores
- **Trust**: transparency rating, follow-through on commitments, contestability usage, and outcomes

If speed improves while safety, fairness, or trust decline, you are borrowing against the future.

Signature Leadership Lines

- Safety fuels speed
- Trust is the strategy
- Resistance reveals where clarity is missing
- People don't resist change. They resist powerlessness in it.
- Middle managers aren't blockers. They are bridges. Invest in them first.
- Psychological safety isn't comfort. It is permission to think, question and grow.
- Resistance is information. Use it.
- Trust is built when leaders tell the truth about limits and still take responsibility for outcomes.

Reflection

Where are your middle leaders expected to deliver certainty without the clarity and confidence they need first?

What will you change this week to close that gap? Where is resistance showing me what is unclear or unsafe?

Which groups need reassurance or involvement before they can move confidently?

What would increase clarity, fairness, or emotional safety this week?

Leadership Closing Thoughts

Leading people through uncertainty requires presence, clarity, and deep respect for the emotional landscape of change. When leaders treat resistance as information, equip middle managers and create psychological safety, they transform fear into capability and hesitation into momentum. In the next chapter, we turn to governance and ethics, exploring how leaders protect trust and integrity as AI becomes more deeply embedded across the workplace.

If change is the journey, culture is the terrain.

Chapter 7:

Culture That Holds During Transformation

Culture is not what you preserve around technology change. It is the foundation that decides whether the change works. AI can scale speed and capacity, but it cannot create shared purpose, values, and human connection. Those are leadership work. This chapter shows how to maintain cultural cohesion while AI reshapes work.

Why this matters

- Strong cultures land complex changes because trust carries the load that technology alone cannot.
- Most AI programs stall on culture, not on code.
- If systems reward what contradicts your values, trust erodes, even when metrics improve.

Values-aligned AI:

translate principles into system behaviour

Stated values must become design requirements. Treat values as specifications for system behaviour, not slogans.

Use this **Values Translation Framework**:

1. **Articulate values in behavioural terms**

   Move from "we value integrity" to decision rules such as "we prioritise transparency in decisions even when it adds effort."

2. **Technical translation**

   Convert rules into design constraints and governance: what data to weight, thresholds for human review, contestability routes, and how the system handles exceptions.

3. **Alignment verification**

   Monitor outputs and impacts, run periodic audits, and review unintended effects with cross-functional oversight.

Practical moves

- **Values impact assessment** before build, not after go-live.
- **Inclusive design teams** that pair engineers with people who hold the cultural story.
- **Tiered explanation protocols** so leaders, operators, and affected employees each get the level of clarity they need.
- **Decision rights map** that states what remains human, what is shared, and what is automated.
- **Plain-language model cards for consequential use cases** so people can see what the system considered and how to contest outcomes.

Cultural continuity: design for human connection

Digital efficiency can dilute relationships if you let it. Protect the human fabric on purpose.

Map connection points

Identify where human interaction creates value beyond transaction: cross-functional problem solving, mentoring, sensitive customer moments, after-action reviews.

Create intentional human moments

- **Connection rituals**: weekly no-tech collaboration sessions for complex issues.
- **Digital boundaries**: what belongs in chat or workflow tools, and what requires a call or room.
- **Space that signals culture**: physical and virtual spaces that encourage spontaneous exchange and reflection.
- **Story cadence**: share short wins where human judgment improved an AI outcome.

Run a connection audit

Before rollout, list current touchpoints, what the tool will change, what must be preserved, and what new rituals will replace what is lost. Document who owns each ritual and how you will know it is working.

Intentional culture evolution: adapt without losing identity

Culture is living. Treat it as adaptively resilient.

The Cultural Evolution Framework

1. **Identity anchors**

   Name the cultural elements that define who you are, not just how you operate. Keep these non-negotiable.

2. **Adaptive pathways**

   Describe how each anchor shows up in an AI-enabled flow. Example: if collaboration is an anchor, build it into system prompts, shared metrics, and review rituals, not posters. Example:

If transparency is an anchor, publish a quarterly "you said, we did" change log tied to AI-influenced decisions.

3. **Evolutionary governance**
   Set a cadence for culture checks: pulse surveys, focus groups, metric reviews, and a "*you said, we did*" loop that publishes changes.

Signals you are on track

- Efficiency rises without drops in collaboration or fairness.
- People can explain how the system reflects values.
- Overrides and exceptions are discussed as learning, not failure.

Two-week starter plan

Week 1

- Run a values impact workshop for the target workflow.
- Draft decision rights and tiered explanations.
- Map connection points and set two new rituals.
- Baseline culture and trust pulses.

Week 2

- Pilot with one team.
- Hold daily ten-minute "exceptions and insights" huddles.
- Publish a short "*you said, we did*" at the end of the week.
- Decide to scale, iterate, or pause based on **culture and performance signals captured in the pilot**.

Measures that matter

Track both what you do and who you become.

- **Performance**: cycle time, accuracy, exception quality, override outcomes.

- **People**: psychological safety, perceived fairness, sense of meaning, manager enablement.
- **Culture**: collaboration index, cross-team help rate, integrity of decisions against values.
- **Trust**: transparency rating, contestability usage and outcomes, follow-through on commitments. Perceived explainability score (frontline rating of "I understand what influenced this decision").

If speed improves while connection, fairness, or trust decline, you are creating cultural debt.

Signature Leadership Lines

- Transparency scales trust.
- Efficiency cannot replace culture, but it can erode it.
- Values do not live in posters; they live in systems.
- Culture that adapts with intention becomes culture that endures.
- What you reward in code is what people learn to value.
- Technology scales capability, culture scales trust.

Reflection

Which parts of your culture are non-negotiable, and where do your current AI decisions visibly protect them?

What ritual or rule will you add this month to make that protection real?

Where has efficiency unintentionally reduced human connection, and what ritual will you re-introduce to restore it?

Closing

Transformation that lasts aligns systems to values, preserves human connection, and evolves culture with intent. Build that alignment into design, not cleanup. Protect the moments where people make meaning. Govern evolution on purpose. Do this and the technology strengthens who you are instead of rewriting it.

Culture creates the conditions; measurement tells you whether those conditions are creating sustainable performance.

Part IV | SCALING WITH INTEGRITY

Chapter 8:

Measuring what matters

(Speed and Adaptability)

In the rush to implement AI, most teams measure what is easy: cost, speed, error rates. The organisations that win measure what matters: decision quality, human experience, learning velocity, and the capacity to adapt when conditions change. Metrics shape behaviour. If you only track efficiency, you will optimise for speed and quietly erode judgment, trust, and resilience.

This chapter offers a practical way to build a measurement system that captures both speed and adaptability, linking technical performance to human outcomes and continuous improvement.

Why this matters

- Traditional dashboards assume stable processes. AI changes the process itself.
- Efficiency gains can mask declining decision quality, employee confidence, or customer trust.

- Balanced measurement prevents "fast but fragile" systems and builds "fast and flexible" ones.
- Balanced measurement prevents fast-but-fragile systems and builds fast-and-flexible ones.

The Dual Horizon Framework

Measure the present and prepare for the future.

Horizon 1: Speed (Run Metrics)

Cycle time, throughput, cost per outcome, right-first-time rates, automation coverage, variance.

Horizon 2: Adaptability (Change Metrics)

Time to reconfigure models and workflows, quality of human overrides, success rate on novel cases, learning adoption (how quickly updates improve outcomes), psychological safety and confidence to use AI, contestability usage and outcomes, perceived explainability, and clarity of decision rights.

Guiding rule

If a speed metric improves while an adaptability metric declines, you are building fragility.

The Hybrid Scorecard

Combine quantitative metrics with qualitative insight across five domains.

1. **Operational Performance**
   - Quant: cycle time, error rate, exception volume, rework.
   - Qual: friction points from user feedback, pattern notes on where errors occur.

2. **Decision Quality**
   - Quant: accuracy against expert review, outcome uplift vs baseline, cost of false positives/negatives.
   - Qual: expert panels' narrative assessments, post-decision debriefs on nuance missed or added.
   - Qual: "second-order effects" notes where a correct decision still produced poor experience.

3. **Human Experience**
   - Quant: psychological safety index, decision confidence, adoption and active use, time in high-value tasks.
   - Qual: stories where human judgment improved or corrected AI, sentiment from frontline forums.

4. **Customer/Stakeholder Outcomes**
   - Quant: satisfaction, resolution quality, repeat contact rate, complaint themes closed.
   - Qual: verbatim feedback samples tied to specific workflow changes.

5. **Integrity & Risk**
   - Quant: explainability coverage, contestability rate and overturn rate, bias drift alerts, audit completion on time.
   - Qual: ethics review notes, "near miss" narratives and mitigations.

Feedback Loops that Make Systems Smarter

Treat feedback as your operating system.

Performance loop

Dashboards for speed, quality, stability. Daily or weekly. Owner: product/ops.

Impact loop

Are we achieving the business and customer outcomes we intended? Monthly. Owner: business unit.

Adaptation loop

How quickly did we respond to change or edge cases? What changed in the model, guardrails, or workflow as a result? Fortnightly to monthly. Owner: cross-functional AI governance with product and risk.

Learning loop

Capture human overrides, contestations, and exception rationales as training fuel. Close the loop by showing what changed. Owner: data/ Machine Learning (ML) with HR and ops.

Practical metrics menu (pick 8–12 max)

Speed

- End-to-end cycle time
- Right-first time %
- Cost per case
- Automation coverage %

Quality

- Expert-review accuracy
- Outcome uplift vs baseline
- Cost of misclassification

Adaptability

- Days to reconfigure after policy change
- Novel-case success rate
- Override effectiveness (rate at which human overrides improve outcomes)
- Time from insight to deployed improvement

Human

- Psychological safety score
- Decision confidence index
- % time in high-value work
- Adoption and active use rates

Integrity

- Explainability availability (who can see what, when)
- Contestability usage and overturn %
- Bias drift alerts addressed on time

Cadence and roles

- **Daily/weekly**: operational huddles on speed and quality, with a 10-minute "exceptions and insights" review.
- **Fortnightly**: adaptation review: novel cases, overrides, contestations, what changed.
- **Monthly**: outcome and integrity councils: decision quality, customer impact, fairness, explainability, bias drift.
- **Quarterly**: scorecard refresh: keep what predicts outcomes, drop vanity metrics, add new signals.

Name clear owners for each metric and each loop. Publish "you said, we did" updates so people see feedback turn into change. Close the loop within two sprints so people see learning become standard.

Two-week starter plan

Week 1

- Identify one workflow to pilot.
- Select 4 speed and 4 adaptability metrics.
- Stand up the daily "exceptions and insights" huddle.
- Add a one-click contest/override path in the tool.
- Baseline decision confidence and psychological safety.
- Capture a 5-question pulse on fairness, explainability and confidence.

Week 2

- Begin fortnightly adaptation review.
- Ship one improvement driven by human feedback.
- Share a short note: what changed, why, and the impact.
- Decide to scale, iterate, or pause.

Red flags to act on

- Speed improves while overrides, complaints, or redo work rise.
- People stop contesting decisions. Silence is not trust. It is fatigue or fear.
- Check for workload or psychological safety issues before assuming satisfaction.
- Model updates are slow, undocumented, or do not move the metrics they targeted.
- Customer satisfaction or frontline confidence dips as throughput climbs.

Signature Leadership Lines

- You manage what you measure.
- If you measure only speed, you will optimise yourself into fragility.
- Efficiency wins headlines. Adaptability wins the future.
- AI accelerates. Leaders ensure direction.
- Data is not neutral. Your metrics teach people what to value.
- Sustainable transformation measures learning, not just output.

Reflection

Which two adaptability metrics, if improved this quarter, will most increase resilience without sacrificing today's performance?

What will you stop measuring to make space for them?

What narrative will you use to retire two vanity metrics this month?

Closing

Measurement is strategy in action. Choose metrics that reward good judgment, protect integrity, and increase your organisation's ability to adapt. Balance speed with adaptability, numbers with narratives, and performance with learning. Do this and your AI program will become not just faster, but stronger.

Once you understand what matters, scaling becomes a question of pacing and prioritisation.

Chapter 9:

Scaling With Integrity, Pace & Prioritisation

The journey from pilot to enterprise scale is the most treacherous stretch of any AI program. Many initiatives stall here, not because the models fail, but because organisations try to sprint without the foundations that let people, processes, and governance keep pace. Scaling well means balancing speed with purpose, ambition with integrity, and innovation with the human capacity for change.

This chapter gives leaders a practical compass: how to decide **where to accelerate**, **where to proceed methodically**, and **how to keep ethics, trust, and value creation intact** while you scale.

Why this matters

Scaling multiplies whatever exists – value or harm. Integrity ensures you scale the right things.

Why scaling stalls

- **Technology treated as the project**. Scaling is an organisational transformation, not a tooling rollout.
- **Speed without readiness**. Teams outpace data, process, and capability maturity.

- **Ethics as afterthought**. Governance is bolted on late, creating "ethics debt."
- **Use-case sprawl**. Too many initiatives dilute focus, budget, and credibility.

Leadership line: Scaling is not acceleration. Scaling is amplification. Only amplify what works and aligns.

Integrity by design: guardrails that scale

Ethics is not a brake. It is the chassis. Build these three guardrails from day one and extend them as you scale:

1. **Principle-based governance**

   Replace rigid rulebooks with clear principles that travel across use cases:

   - **Purpose boundaries**: what AI will and will not be used for.
   - **Decision rights**: when AI can decide, when humans must review, and how to escalate.
   - **Transparency standards**: explanation depth that scales with consequence. Include a plain-language rationale for consequential decisions visible to affected people.
   - **Data stewardship**: consent, quality, lineage, and retention expectations.

2. **Distributed accountability**

   Move beyond a lone central committee. Embed responsibility in the work:

   - Ethics champions in each team.
   - Lightweight checklists and decision templates.
   - Clear escalation paths and consequence management.
   - A visible "you said, we did" log of issues raised and resolved.
   - A one-page "decision rights" sheet to accompany every new use case.

3. **Continuous assessment**
   Integrity is a process, not a state:
   - Algorithmic impact assessments before launch.
   - Bias drift and performance monitoring after launch.
   - Contestability channels with overturn tracking and documented remediation.
   - Quarterly ethics reviews focused on patterns, not isolated incidents.

Leadership line: Integrity is not a cost. It is insurance against fragility and reputational loss.

The pace paradox: finding the right speed

You do not need one speed. You need **pace layering**.

- **Fast lanes** (learn by doing): sandboxes, low-risk automation, internal-facing tools, data pipelines.
- **Deliberate lanes** (trust by design): external-facing decisions, high-stakes HR or customer impacts, regulated domains.
- **Enablement lane** (always on): capability building for middle managers, frontline coaching, change comms, and playbooks.

Embed an enablement budget line so capability and change comms never become optional.

Red flags

- Hitting technical milestones while adoption lags.
- Silence in contestation channels.
- Rework climbs as throughput rises.
- "Temporary" exceptions become the norm.

Leadership line: Move at the speed of trust. Accelerate where trust is earned and risks are bounded.

Prioritisation that focuses value

When everything is a priority, nothing scales. Use two simple lenses together.

1. **The Value–Readiness Matrix**

   Plot each initiative on **business value** (impact) and **implementation readiness** (data, process, people, governance).

   - **Quick wins** (High–High): ship first to build momentum.
   - **Strategic bets** (High value–Lower readiness): phase delivery, invest in data and capability while releasing increments.
   - **Foundation builders** (Lower value–High readiness): enable future wins (data quality, MLOps, model registry, access controls).
   - **Reconsider** (Lower–Lower): pause unless a clear dependency exists.

   Re-score monthly. If an item sits unchanged for two cycles, retire or re-scope.

2. **The Human–AI Value Lens**

   Select use cases that **enhance** people, not just replace tasks:

   - **Efficiency value**: time back from routine work.
   - **Decision value**: improved quality and consistency.
   - **Experience value**: better customer and employee journeys.
   - **Capability value**: skills, judgment, and learning that persist beyond this tool.

Leadership line: Prioritise where AI makes your people better at the work only humans are best placed to do.

Pacers and accelerators: how to sequence scale

Create a one-page decision rubric. For each use case, score 1–5 on:

- **Impact**: revenue, cost, risk, or service uplift.
- **Readiness**: data quality, process clarity, integration.

- **Risk**: consequence of error, compliance exposure.
- **Adoption likelihood**: stakeholder appetite, middle-manager enablement, training.
- **Learning yield**: what this teaches that unlocks later value. Example: A moderate-impact customer triage pilot with high override learning can unlock three later use cases.

Decision rules

- High impact + high readiness → **Accelerate**.
- High impact + medium readiness + medium risk → **Phase** with guardrails and early benefits.
- Medium impact + high learning yield → **Pilot** as a capability builder.
- Low impact + low learning yield → **Defer**.

Operating model for scale

Roles that matter

- **Product owner** (business): value case, outcomes, adoption.
- **ML lead** (data/tech): model performance, drift, and resilience.
- **Process owner** (operations): workflow design, handoffs, SOPs.
- **Risk & ethics partner**: guardrails, contestability, impact reviews.
- **Change lead**: comms, training, feedback loops, middle-manager enablement.

Cadence

- Maintain a visible portfolio wall with status, owners, and next decision dates.
- **Weekly**: delivery stand-ups; exceptions and insights.
- **Fortnightly**: adoption and adaptation review; overrides, contestations, novel cases.
- **Monthly**: value and integrity council; outcome trends, bias drift, audit actions.
- **Quarterly**: portfolio reset; re-rank, reallocate, retire, or scale.

90-day scaling playbook

Days 1–30: Prove and prepare

- Select two quick wins and one foundation builder.
- Baseline value, adoption, trust, and risk indicators.
- Stand up contestability and a visible "changes shipped" log.
- Train middle managers on narrative, handoffs, and coaching.

Days 31–60: Phase and learn

- Ship increments every 2–3 weeks with before/after metrics.
- Capture human overrides as training fuel and close the loop.
- Run an algorithmic impact assessment for the next high-impact use case.
- Publish a short win's digest: what changed, why, impact.

Days 61–90: Scale and safeguard

- Promote the best quick win to two adjacent teams.
- Lock in playbooks and templates; reduce bespoke effort.
- Refresh the portfolio using the Value–Readiness Matrix.
- Retire one low-value initiative to signal focus. Publicly explain the retirement to model disciplined focus.

What to measure while scaling

Pull from Chapter 8's hybrid scorecard and add scale signals:

- **Replication success**: time to onboard a second team to the same use case.
- **Change velocity**: time from feedback to deployed improvement.
- **Risk posture**: explainability coverage, contestations and overturns, bias alerts closed on time.
- **Human uplift**: decision confidence, psychological safety, time in high-value work.
- **Value realisation**: outcome uplift sustained beyond the pilot window.

Leadership line: If scaling increases tickets and decreases trust, you are multiplying problems, not value.

Common failure modes (and fixes)

- **Use-case sprawl** → Create a portfolio wall. Anything not on the wall does not get talent or money.
- **Shadow AI** → Offer a safe path to register experiments. Reward disclosure, not punishment.
- **Model drift ignored** → Automate drift alerts and tie them to change SLAs.
- **Manager squeeze** → Provide middle-manager toolkits, talking points, and coaching. Measure their confidence, not just frontline adoption.
- **Ethics debt** → Time-box go-live on completion of impact assessment and explainability check.

Signature Leadership Lines

- If scaling increases tickets and decreases trust, you are multiplying problems, not value.
- When leaders chase speed without readiness, they do not scale. They scatter.
- You can deploy AI everywhere fast or deploy it where it matters and win.
- Governance is how values travel at scale.
- The portfolio you kill signals more discipline than the portfolio you fund.

Reflection

Where are you scaling because it is strategic, and where are you scaling because it feels urgent?

Which single use case will you retire to free capacity for the next strategic bet?

What will you pause this quarter to free capacity for what matters most?

Closing

Scaling with integrity is a choice. Choose principles that travel, a pace that compounds trust, and priorities that amplify what already works. Do this, and scale becomes more than bigger or faster. It becomes better.

Scaling well builds maturity; futureproofing turns maturity into momentum.

Chapter 10:

Futureproofing Your Organisation

Leadership Strategies for the Evolving AI-Human Workplace (Future-Ready Leadership & Stewardship)

The organisations that will thrive are not the ones that perfectly predict AI's path. They are the ones that build the adaptive capacity to evolve with it while staying anchored to a human core. Future-proofing shifts leadership from static plans to strategic resilience: sensing change early, interpreting it wisely, and responding with agility grounded in values and purpose.

Why This Matters

The future belongs to leaders who evolve with technology while staying anchored in human values.

This chapter offers three capabilities leaders can put to work now:

1. **Anticipatory leadership** to see what is emerging and convert signals into options.

2. **Organisational resilience** to adapt structures, talent, and processes without exhausting people.
3. **Ethical stewardship** to ensure technology serves human wellbeing and trust at scale.

Leadership line: The winning move is not prediction. It is preparation with purpose.

1. Anticipatory leadership: sensing, interpreting, acting

Reactive leadership waits for certainty. Anticipatory leadership builds it.

The Anticipatory Leadership Framework

- **Horizon scanning**: Cross-functional teams track defined AI domains (e.g., GenAI for service, risk modelling, safety). Each quarter, they publish two-page briefs: capability snapshots, relevance to our context, competitor movements, and "no-regrets" experiments.
- **Implication workshops**: Translate signals into consequences for your customers, cost base, risk profile, operating model, and brand. Explore upside, downside, and second-order effects.
- **Option portfolios**: For each signal, create three options: **pilot now, prepare to scale, monitor**. Assign owners, criteria to advance, and decision dates. Kill or advance options on a schedule. Stale options drain attention and credibility.
- **Learning cadence**: After every pilot, publish "what we learned" within 10 days, including what to stop, start, and standardise.

Leadership line: Don't ask "What will AI do to us?" Ask "What options do we want to hold when it arrives?"

Reflection prompt: Which two AI signals deserve an option this quarter, and who owns advancing them?

2. Organisational resilience: structures that can keep evolving

One-time transformations fail in environments that keep changing. Build for continuous adaptation.

Four building blocks

A. **Modular design**

Break big systems into components with clear interfaces. Update parts without rebuilding the whole.

Actions: Map core capabilities; define handoffs and data contracts; set change SLAs per module.

B. **Capability networks, not rigid hierarchies**

Mobilise talent into configurable pods around problems and skills (data, product, risk, experience). Publish a skills inventory and a visible internal marketplace so people can move to priority pods.

Actions: Stand up a skills inventory; create a bench of "swing" roles; fund pods from a portfolio, not a single cost centre.

C. **Decision rights that fit hybrid work**

Clarify who decides, when human review is mandatory, and how to escalate.

Actions: Publish RACI for each AI use case; require a plain-language rationale for consequential decisions; track overrides and overturns.

D. **Institutionalised learning**

Make adaptation visible and repeatable.

Actions: Run monthly adaptation reviews on drift, exceptions, and user feedback; promote the best fix into a standard within two sprints.

Leadership line: Resilience is not rigidity or chaos. It is modularity with clear rules of engagement.

Reflection prompt: Which one module can you redesign now to make the next upgrade 50 percent easier?

3. Ethical stewardship:
responsible innovation that scales

Ethics is how your values travel through your systems. Stewardship enables speed with trust.

The Ethical Innovation Framework

Values alignment

Translate values into design requirements. Example: If inclusion matters, define how models weigh factors, how explanations are shown, and how accessibility is handled.

Ethical risk assessment

Assess stakeholder impacts before build. Consider employees, customers, and communities. Decide on mitigations and red lines upfront.

Transparency and contestability

Explain what the system considered and how to challenge outcomes. Track contests, overturn rates, and learn from them.

Ongoing monitoring

Watch for bias and drift, not just accuracy. Set thresholds that trigger retraining, rollback, or human-only operation. Define who triggers rollback, how long rollback can persist, and what re-entry criteria apply.

Governance that empowers

Use principle-based guardrails, distributed ethics champions, and a clear escalation path. Publish a "you said, we did" log to reinforce trust.

Leadership line: Technology scales capability. Stewardship scales trust.

Reflection prompt: Where is the greatest gap between our stated values and how our systems currently behave?

Putting it together:
a 120-day future-readiness sprint

Days 1–30: Sense and align

- Stand up two horizon scanning cells and agree on signal domains.
- Run one implication workshop; create three option briefs.
- Draft values-to-design criteria for the next high-impact use case.

Days 31–60: Pilot and protect

- Launch one no-regrets pilot with a plain-language model card and contestability live on day one.
- Map one function into modules; define interfaces and decision rights.
- Train a pod of middle managers on hybrid decision coaching and change narratives. Include practice on contestability and coaching after overrides.

Days 61–90: Learn and lock

- Publish pilot learnings; promote two practices to standards.
- Implement bias and drift monitoring; set rollback triggers.
- Formalise a cross-functional adaptation review cadence.

Days 91–120: Scale and sustain

- Replicate the winning pilot in a second team; measure replication time.
- Launch an ethics champions network; publish the first "you said, we did."
- Refresh the option portfolio; retire one low-value initiative to fund the next bet.

What to measure:
the future-ready scorecard

- **Option health**: number of live options with defined criteria to advance.
- **Replication time**: weeks to roll a proven use case to a second team.
- **Adaptation velocity**: time from feedback to shipped improvement.
- **Trust signals**: explanation coverage, contest rates, overturn rates, issue closure time.
- **Human uplift**: decision confidence, psychological safety, time in high-value work.
- **Values alignment**: audit variance between stated principles and system behaviour. Run a quarterly audit and publish changes taken.

Leadership line: Measure learning and integrity with the same discipline you apply to speed.

Signature Leadership Lines

- The winning move is not prediction. It is preparation with purpose.
- Adaptability is the new advantage.
- Stewardship scales trust.

Reflection

Where is the greatest gap between our stated values and how our systems currently behave?

Which two AI signals deserve an option this quarter, and who owns advancing them?

Which single module could you redesign now to make the next upgrade 50 percent easier?

Closing

Locally, Australia's federal policy direction adds weight to these measures. Speed and efficiency alone are no longer considered sufficient indicators of AI success. Decision quality, fairness, safety and trust are now recognised as critical signals of readiness and performance.

Futureproofing is a leadership discipline. Sense early. Interpret with context. Act with integrity. Design structures that can change, and cultures that want to. When you do, AI stops being a wave to withstand and becomes a current you can navigate toward meaningful work, better decisions, and durable advantage. Future-ready leadership is a cadence: sense, interpret, act, learn, and align.

Conclusion

The Path Forward

The workplace we've known for generations is undergoing its most profound transformation since the industrial revolution. As artificial intelligence moves from experimental technology to essential business partner, we stand at a pivotal moment that will redefine not only how work gets done, but what it means to be human at work. This is a leadership era where technology is a co-worker and trust is the operating system.

Throughout this book, we've explored the multifaceted journey of integrating AI into organisations: assessing readiness, designing hybrid teams, building governance frameworks, maintaining culture through transformation, measuring what truly matters, and scaling with integrity. At every stage, one truth has remained constant: the most successful AI implementations are those that enhance, not diminish, our humanity.

As we close this exploration, three leadership principles rise above all others. They form the compass for navigating the AI-enabled workplace with confidence and care.

1. Lead with purpose, not just capability

The central question is no longer "*What can this technology do?*" but "*What do we need/want this technology to do to advance our purpose?*"

The organisations that thrive will be those that harness AI in service of meaningful human objectives, not those that adopt technology for its own sake. Purpose-driven leaders start with clarity: how each innovation will strengthen their mission, improve the experience of customers, and create work that matters for their people.

Leading with purpose also requires courage: the courage to decline opportunities that distract from your values, even when competitors race ahead. Use a simple test: Does this advance our purpose, improve the work our people do, and honour our values?

Purpose provides both speed and stability. It anchors transformation in meaning.

2. Prioritise trust over efficiency

Efficiency may open the door to innovation, but trust keeps it open.

The metrics that dominate many AI conversations: cost, speed, accuracy; are important, but incomplete. The true measure of success is whether AI strengthens trust: trust between the organisation and its customers, between leaders and teams, and between humans and the systems that support their decisions.

Trust grows through transparency, clear accountability, and alignment between words and actions. It means being honest about what AI can and cannot do. It means designing for contestability, allowing people to question, override, and improve the system, because accountability cannot be automated. Make contestability visible and safe so people can improve the system without fear.

Organisations that chase efficiency at the cost of trust might gain temporary speed, but they lose long-term traction. Broken trust is far costlier than a delayed project. Sustained performance depends on confidence that technology amplifies human judgment rather than replacing it.

3. Invest in human potential alongside technology

The ultimate differentiator in the AI era will not be the sophistication of algorithms but the adaptability of people. Future-ready organisations invest as heavily in developing human capability as they do in digital infrastructure.

That means building learning ecosystems that grow uniquely human strengths: ethical reasoning, creative problem-solving, empathy, and collaboration. It means designing roles that let people do what machines cannot: connecting ideas, interpreting context, and exercising care. Prioritise capability that compounds: ethical judgment, creative problem solving, collaboration, and sense-making.

Technology best used will liberate human energy, not constrain it. The purpose of AI is not to replace judgment, but to create the conditions where judgment matters more.

The road ahead

As you move forward, remember that implementing AI is not the destination. It is the beginning of a new form of leadership that blends technological literacy with human insight, and strategic ambition with moral clarity.

You are not simply building systems; you are shaping the future of work. Every choice you make about how AI is deployed, governed, and communicated teaches your organisation what it values. Every design decision is also a cultural one.

The future of the workplace will belong to leaders who combine intelligence with empathy, innovation with integrity, and progress with purpose. Those who see AI not as a replacement for humanity, but as a catalyst for human potential.

Begin with one workflow, one governance rhythm, and one learning loop you can sustain.

Technology will keep evolving. What endures is our capacity to learn, to lead, and to care. **That is how progress stays human**.

Appendices

Appendix A: Chapter Checklists

✓ Chapter 1: The New Work Equation

- Define where human strengths must remain central
- Map where trust is fragile
- Align leadership narrative
- Identify roles most affected
- Clarify expectations early

✓ Chapter 2: Readiness

- Conduct leadership, culture and capability scans
- Align decision rights
- Establish guardrails
- Select pilot areas
- Publish readiness summary

✓ Chapter 3: Hybrid Teams

- Map work to human/AI/shared
- Define hybrid roles
- Establish handoff protocols
- Build override pathways
- Stand up learning loops

✓ Chapter 4: Governance

- Create AI Governance Committee
- Develop RACI-E per use case
- Define transparency standards
- Run impact assessments
- Set monitoring cadence

✓ Chapter 5: Capability

- Baseline capability and confidence
- Define priority human skills
- Build learning ecosystem
- Support psychological safety
- Track confidence signals

✓ Chapter 6: Leading Through Change

- Equip middle managers
- Build challenge channels
- Establish rituals (exceptions & insights)
- Stand up Manager Experience Lab
- Publish feedback loops

✓ Chapter 7: Culture

- Translate values into system design rules
- Run connection audit
- Align rituals and boundaries
- Baseline culture and trust
- Set evolution cadence

✓ Chapter 8: Measuring What Matters

- Select 8–12 metrics
- Stand up daily/weekly huddles
- Build contestability path
- Run adaptation reviews
- Refresh dashboard quarterly

✓ Chapter 9: Scaling With Integrity

- Complete Value–Readiness Matrix
- Apply pace layering
- Build decision rubric
- Stand up integrity monitoring
- Retire one initiative before adding another

✓ Chapter 10: Future-Proofing

- Stand up horizon scanning cells
- Run implication workshop
- Define option portfolio
- Begin modular redesign
- Track future-ready scorecard

Appendix B: 120-Day Organisational Roadmap

Days 1–30: Sense & Align

- Readiness audit
- Middle-manager enablement
- Governance principles
- Contestability live
- Pilot workflow selected

Days 31–60: Pilot & Learn

- Live pilot with human review
- Capture overrides
- Publish learnings
- Begin adaptation cadence
- Ship small improvements

Days 61–90: Scale What Works

- Expand pilot to one adjacent team
- Standardise rituals
- Refresh Value–Readiness Matrix
- Lock in operating model for scale
- Strengthen monitoring

Days 91–120: Future-Readiness

- Launch ethics champions
- Modular redesign for one function
- Update option portfolio
- Publish scorecard
- Retire one low-value initiative to fund learning

Appendix C: Australian Policy Lens

Australian AI Landscape: Leadership Essentials for Today's Organisations

Australia's 2025 National AI Plan, released in the week before this book was published, establishes a national direction for responsible, inclusive and secure AI adoption. The plan highlights three commitments that shape how organisations across the economy should prepare.

1. Capture the Opportunities

Focus areas include:

- Responsible innovation
- Local capability
- Sovereign data and compute infrastructure
- Industry and research partnerships

2. Spread the Benefits

Emphasis includes:

- Worker capability and inclusion
- Support for SMEs
- Digital equity across regions and demographics
- Skills pipelines and lifelong learning

3. Keep Australians Safe

Commitments include:

- Strong governance expectations
- Transparency and documentation
- Updated guardrails for privacy and consumer protection
- The establishment of the AI Safety Institute
- Clear accountability for AI-influenced decisions

Implications for Leaders

Leaders are increasingly expected to:

- Strengthen governance maturity
- Equip workers with confidence and capability
- Support transparency and contestability
- Embed psychological safety
- Measure trust alongside performance metrics

Appendix D: Glossary

A

Adaptive capacity
The organisational ability to adjust structures, processes and behaviours in response to changing conditions.

AI governance
The principles, structures and routines that ensure AI systems behave in line with organisational values, risk appetite and legal obligations.

AI maturity
The level of capability, integration and confidence an organisation has in effectively using AI across workflows.

Algorithmic decision-making
Decisions made or influenced by AI systems using data-driven models.

Algorithmic impact assessment (AIA)
A structured process that evaluates risks, benefits and fairness implications before an AI system is deployed.

Automation coverage
The percentage of tasks or steps in a workflow that are fully or partially automated.

C

Capability architecture
The design of human strengths, skills and workflows that complement AI's role in an organisation.

Change fatigue
Exhaustion caused by repeated or poorly managed change.

Contestability
A mechanism allowing people to question, appeal or override AI-influenced decisions.

Cultural debt
The erosion of trust, values or connection caused when systems are scaled faster than culture can sustain.

Cycle time
The total time it takes to complete a process or case from start to finish.

D

Data lineage
Documentation of where data originated, how it has been used and how it has changed.

Decision rights
Clear rules outlining who makes which decisions, when AI can act, and when human review is mandatory.

Digital twin (workforce)
A virtual simulation used to model workflows, change impacts and resource allocation.

Drift (model drift)
When an AI system's performance declines because data or conditions have changed.

E

Explainability
The ability to describe how an AI system reached a decision in language appropriate to leaders, employees and customers.

Exception handling
Processes used when AI produces uncertain, incorrect or incomplete results, requiring human judgment.

Ethics champions
Selected employees who promote responsible AI practices within their team or function.

F

Feature importance

A description of which data points most influenced an AI model's decision.

Feedback loop

A structured mechanism that uses human or system input to improve models, workflows or behaviours.

H

Human-in-the-loop

A workflow design where humans maintain oversight, context, judgment and ultimate accountability for decisions informed or produced by AI.

Hybrid workflow

A process where humans and AI each play defined roles to deliver a combined result neither could produce alone.

Horizon scanning

Systematic tracking of emerging trends, technologies and risks to inform strategy.

I

Impact audit

A review of how AI is affecting outcomes, fairness, customer experience or employee confidence.

In-the-flow learning

Learning embedded in daily work rather than in formal training sessions.

Integrity signals

Leading indicators of ethical performance, such as contestation rates, overturned decisions or bias alerts.

L

Learning velocity

The pace at which insights from feedback, overrides or errors improve system performance or workflow design.

Leadership narrative

A clear, consistent explanation from leaders about why AI is being used and what it means for people.

M

Meaningful human control

A requirement that humans can understand, intervene in and override system decisions where needed.

MLOps (Machine Learning Operations)

The discipline combining machine learning, engineering, risk and operational practices to manage AI models in production.

Model card

A plain-language summary explaining what an AI model does, what data it was trained on, its limits and its appropriate use cases.

O

Override

A human decision to supersede or correct an AI recommendation.

Operational resilience

An organisation's ability to absorb disruption and maintain performance.

Option portfolio

A set of pilots, experiments and "no-regrets" actions that keep the organisation adaptable.

P

Pace layering

Designing workstreams to move at different speeds depending on risk, trust and readiness.

Performance drift

A decline in the consistency or accuracy of AI outputs due to changing context.

Predictive signal

A data point that strongly correlates with a future trend or outcome.

R

RACI-E

A decision-ownership model defining who is Responsible, Accountable, Consulted, Informed and who manages Escalations.

Readiness audit

A structured assessment of leadership alignment, cultural maturity, capability, systems and risk posture before adopting AI.

Risk posture

An organisation's tolerance for uncertainty and potential harm in pursuit of innovation.

S

Sandbox

A safe, limited environment where teams can test AI systems without impact on real operations.

Shadow AI

AI tools adopted by individuals or teams without governance oversight, often creating hidden risks.

Skill adjacency

Capabilities closely related to existing strengths that can be developed to open new roles or opportunities.

Systemic trust
Trust that results from consistent behaviours, clear guardrails and transparent systems rather than individual relationships alone.

T

Tiered explanation
Different levels of detail provided depending on the audience (technical, managerial, customer/employee).

Trust pulse
A short assessment measuring employee confidence, safety and perceived fairness during transformation.

V

Values alignment
Ensuring AI design and impacts reflect the organisation's stated values.

Value–Readiness Matrix
A tool for prioritising AI use cases by comparing their potential impact with the organisation's ability to implement them.

W

Workflow orchestration
Coordinating how humans and AI interact across steps in a workflow to achieve a shared outcome.

Workforce modularity
Structuring teams and capabilities so they can be reconfigured quickly as needs and technologies evolve.

Further Reading

The topics explored in *The AI-Enabled Leader* intersect with many disciplines, from organisational psychology and people strategy to technology ethics and change leadership. The following works provide deeper exploration of some of the foundational ideas introduced in this book.

If you are ready to explore these ideas more deeply, the works below provide valuable next steps. They expand on the concepts introduced in this book and offer practical guidance for leaders who want to strengthen people systems, ethical governance, and human capability in an AI-enabled world.

By the Author

Their Workplace, Your Business: A Founder's HR Playbook for Scalable Success

Barratt-Browne, Beth. *Better HRx Publishing, 2025.*

A practical guide for building people systems that enable organisational growth. This book covers leadership, culture, compliance, and the core disciplines every business needs to establish before layering in technology or scale.

Leadership and Organisational Transformation

The Fearless Organization

Amy C. Edmondson

A seminal text on psychological safety and the conditions that enable teams to learn, innovate, and perform.

Leading Wellbeing: A Leader's Guide to Building Psychological Safety
Fleur Heazlewood

A practical Australian perspective on wellbeing and leadership impact.

Reinventing Organizations
Frederic Laloux

Explores new models of organisational design that empower people and purpose in times of change.

Human-Centred Technology and AI

Human + Machine: Reimagining Work in the Age of AI
Paul R. Daugherty & H. James Wilson

An accessible overview of how AI can enhance rather than replace human capability.

Atlas of AI
Kate Crawford

A powerful exploration of the human, ethical, and environmental implications of artificial intelligence.

Tools and Weapons
Brad Smith & Carol Ann Browne

Insights from Microsoft's leadership on responsible technology and global accountability.

Governance, Ethics and Scaling with Integrity

Competing in the Age of AI
Marco Iansiti & Karim R. Lakhani

Examines how organisations redesign governance and operating models for the AI era.

Measure What Matters
John Doerr

A foundational guide to goal setting and measurement frameworks (OKRs) that translate purpose into action.

The Infinite Game
Simon Sinek

Encourages leaders to prioritise sustainability and integrity over short-term wins, a vital mindset for AI transformation.

For HR and People Leaders

No Rules Rules: Netflix and the Culture of Reinvention
Reed Hastings & Erin Meyer

A study in how culture and trust outperform policy and control.

Work Rules!
Laszlo Bock

Lessons from Google's people operations on building engagement, autonomy, and scalable culture.

AI 2041
Kai-Fu Lee & Chen Qiufan

A forward-looking exploration of how artificial intelligence may shape work, society, and leadership by 2041.

Bibliography

References

The ideas in this book are grounded in contemporary research across leadership, organisational psychology, technology ethics, and AI-enabled work design. The sources listed in the Bibliography directly informed specific arguments, models, or data points referenced throughout the chapters. The Further Reading section provides additional works that deepen exploration of these themes but are not necessarily cited in the text.

AFL-CIO. *Workers First: AI.* AFL-CIO, 2024. https://aflcio.org/reports/workers-first-ai

Bain & Company. "Adapting Your Organization for Responsible AI." January 2024. https://www.bain.com/insights/adapting-your-organization-for-responsible-ai/

Bauknight, P. "Artificial Intelligence for the Safety Professional: Benefits, Risks, and Legal Implications." *ASSP*, March 2025.

BDO. *AI-Enabled Workforce Change Readiness Assessment.* 2023.

Bhatija, S. "OSHA Compliance in the Age of AI." Spot.ai, March 2025.

Bleuel, F. "Leadership in the Age of AI: Precision and Paradox." HPI d-school, 2025.

Brewer, T. "Designing Hybrid Teams: Blending AI with Human Expertise." Functionly, 2025.

Crawford, K. *Atlas of AI.* Yale University Press, 2021.

Daugherty, P., & Wilson, H. *Human + Machine: Reimagining Work in the Age of AI.* Harvard Business Review Press, 2018.

Edmondson, A. C. *The Fearless Organization.* Wiley, 2019.

EY Global. "The Role of Responsible AI Principles in Driving Competitive Advantage." 2023.

Fairwork. "Fairwork AI Principles." 2023–2024.

Future Processing. "AI Readiness Assessment: Are You Prepared for AI Integration?" 2025.

Global AI Policy Frameworks (selected): Australia's National AI Plan (Australian Government, 2025); European Union Artificial Intelligence Act (European Commission, 2024); US NIST AI Risk Management Framework (National Institute of Standards and Technology, 2023); Singapore's Model AI Governance Framework (Infocomm Media Development Authority, 2023); UK AI Regulation White Paper and AI Safety Institute Reports (UK Government, 2023–2024).

Golan, G. "AI Governance Framework: Key Principles & Best Practices." MineOS, 2025.

Heazlewood, F. *Leading Wellbeing*. Major Street Publishing, 2022.

Hougaard, R., & Carter, J. "How AI Can Make Us Better Leaders." Potential Project, 2024.

Iansiti, M., & Lakhani, K. R. *Competing in the Age of AI*. Harvard Business Review Press, 2020.

IBM. "From Principles to Action: Building a Holistic Approach to AI Governance." 2024.

ISO. "Responsible AI Ethics Guidance." 2023.

Laloux, F. *Reinventing Organizations*. Nelson Parker, 2014.

Lee, K-F., & Qiufan, C. *AI 2041*. Currency, 2021.

Mackenzie, R. "AI With Integrity: Reimagining the Future of Leadership." Inspirational Group, 2025.

McKinsey & Company. "Superagency in the Workplace: Empowering People to Unlock AI's Potential." 2025.

Microsoft. "Responsible AI Principles and Approach." 2023.

MineOS. "AI Governance Framework." 2025.

Nemko Digital. "Modern AI Governance Frameworks: 2025 Enterprise Implementation Guide." 2025.

Panorama Consulting Group. "Is Your Business Ready for AI?" 2025.

Palo Alto Networks. "What Is AI Governance?" 2024.

Relyance AI. "The Mathematical Nature of Trust: Building Your Trust Architecture." 2023.

Russell Reynolds Associates. "The Four Steps CEOs Need to Take to Build AI-Powered Organizations." 2025.

SHRM. *The AI-Enabled HR Leader: Driving Workforce Transformation*. 2023.

Sinek, S. *The Infinite Game*. Portfolio, 2019.

Smith, B., & Browne, C. A. *Tools and Weapons.* Penguin Press, 2019.
Team Flow Institute. "The Leadership Paradox: Why Slowing Down Matters in AI." 2023.
Trusted Advisor Associates. "Understanding The Trust Equation." 2000.
Visionify.ai. "Top Workplace Safety Trends to Watch in 2025." 2025.
WitnessAI. "AI Governance: Ensuring Responsible Development." 2025.
Woods, G. *The AI-Driven Leader.* SoBrief, 2024.

About the Author

Beth Barratt-Browne is a best-selling author and seasoned HR and business executive with over 20 years leading People & Culture functions across government, commercial, and not-for-profit sectors. As the Founding Principal of Better HRx, a fractional HR consultancy, Beth brings executive-level people leadership to growing businesses without the full-time overhead.

Beth's deep experience with humans at work uniquely positions her to guide organisations through the complex transition to hybrid AI-human workplaces. Beth combines strategic clarity, operational expertise, and a human-centered approach to help leaders design, recruit, and manage AI-integrated teams where technology amplifies human judgment and culture remains central.

Throughout her career, Beth has helped CEOs, founders, and boards solve complex people challenges from leadership transitions and workforce restructuring to embedding performance cultures and strengthening compliance. Her calm, commercial, and collaborative style is informed by her senior roles including Chief Workforce Officer, GM People, and Commercial Director, as well as her service as a Non-Executive Director and Chair of a People & Culture Committee.

Beth holds an MBA from AGSM and completed the esteemed AICD Company Directors Course, further underscoring her leadership and governance credentials.

As a best-selling author recognised for cutting through complexity with clarity, Beth leverages storytelling and evidence-based insights to equip leaders ready to win in the AI-human hybrid workplace. Through Better HRx, she offers fractional Chief HR (People) Officer support and strategic advisory for purpose-driven businesses poised to scale with confidence and integrity.

Beth is the expert leaders turn to when they want to maximize human potential alongside AI not by replacing people, but by empowering them to thrive in the future of work.

Set your pace and plan your supports -
if Beth can be of help, please reach out
via website or LinkedIn

Beth will be facilitating workshops and
keynotes on the areas covered in this book

Enjoyed this Book?

Please visit our website for more information and
assistance and to leave a review

BetterHRx.com.au

Made in the USA
Monee, IL
05 November 2026

67811473R00075